THE GUIDE OF EVERY PROFESSIONAL OF THE FUTURE

ALL ABOUT
DIGITAL TRANSFORMATION
AND THE MINING 4.0

BRIAN PAJARES

All About Digital transformation and the Mining 4.0

1st Edition
1st Edition

By Brian Pajares

Published by InnovaYap Inc

Jr. John Paul II # 265

Cajamarca, Peru

www.innovayap.com

Copyright © 2021 by InnovaYap, Peru

No part of this publication may be reproduced, stored in a retrieval system, or transmitted in any form whether electronic, mechanical, photocopied, recorded, scanned or otherwise, except as permitted without the prior written permission of the publisher, or the authorization by paying the corresponding fee for copying to Copyright.

To obtain permission, you must contact the Legal Department, InnovaYap, Inc. E-mail: brian.pajares@innovayap.com

About the author

Leader and specialist in the implementation of Projects in Open Innovation, Digital Transformation, Disruptive Technology, Adoption of Innovative Culture and Continuous Improvement. Mechanical Engineer from PUCP, collegiate and specialist in Industrial Automation from TECSUP. I have a Global MBA from CENTRUM Catholic, University of Maastricht (Holland) and University of Victoria (Canada). Master in Project Management from the University of Maryland, USA (Fullbright Scholarship). PMP certificate. Work experience of more than 10 years focused on Project Management, Innovation, Culture, Technology and Continuous Improvement in different industries such as Mining, Agro-industry and Manufacturing. Lecturer, educator and writer on topics of innovation, technology and continuous improvement

Contact me: brian.pajares@innovayap.com

Page: www.innovayap.com

Dedication

Dedicated to my parents Valentina Correa and Arturo Pajares. They taught me that without effort and perseverance nothing is achieved in this life. To my brothers who every day continue to be an example to follow. To Yesenia for all her love. To my family for their unconditional support and to my friends for always teaching me new things.

"The advancement of technology is based on making it fit so that we do not even realize it, and thus make it part of everyday life"

Bill Gates, *Co-founder of Microsoft*

Table of Contents

About the author iii
Dedication iv
Table of Contents v
Introduction 1
Chapter 1: 2
   Industrial Digital Transformation in the XXI Century 2
     Digital Transformation 3
       The pandemic effect 4
       Why is digital transformation essential for all companies? 4
       Strategy for digital transformation 5
       Benefits of implementing a digital transformation strategy 17
Chapter 2: 22
   Success Stories of Digital Transformation Projects 22
     Examples of digital transformation in the industry 23
       Volvo case 23
       Lavihood case 23
       Polaris case 24
       Howden case 24
       Woodward case 25
       Vodafone case 25
     Global digital transformation statistics 26
Chapter 3: 29
   General concepts of Mining 29
     The decline in value of the mining sector 30
       Exploration 30
       Construction 33
       Exploitation 35
       Process Plant 36
       Smelting 37
       Refinement 38
       Commercialization 38
Chapter 4: 39
   Mining 4.0 39

MINING VALUE CHAIN 4.0 41
Mining 4.0: Exploration 42
Digitization and processing of geological data with Artificial Intelligence 42
Drones in exploration 43
IoT in geology 44
Mining 4.0: Construction 45
BIM technology 45
Augmented Reality (AR) and Virtual Reality (VR) in construction 46
Robotics in construction 47
Digita twin in construction 48
Mining 4.0: Exploitation 49
Autonomous Teams 49
Electric vehicles in mining 52
Augmented Reality (AR) and Virtual (VR) in the Farm 54
Drones for Exploitation 54
Predictive maintenance in mobile operating equipment 56
Wearable to improve safety and productivity 57
Mining 4.0: Concentration and process plant 59
Digital twin at the mill 59
Predictive maintenance in plant equipment 60
Internet of Things (IoT) in the process plant 62
Mining 4.0: Smelting and Refining 67
Using robots 68
Better energy efficiency 69
Mining 4.0: Comercialization 70
Blockchain 70
IoT in commercialization 71
Digital Mine Control Center 73
Digital Transformation in the Support Area 74
The Total Digital Mine 75
Chapter 5: 76
Conclusions and Recommendations 76
Conclusions and recommendations for a Mining 4.0 and a successful Digital Transformation strategy 77
Bibliography 80

Introduction

The book "All about Digital Transformation and Mining 4.0", is part of a series of 3 books that allows the reader, even if they are not a specialist, to understand the importance of innovation and digital transformation in the energy mining sector, as well as to know the jobs in mining innovation that are currently being worked on in the different areas of mining, the future that lies ahead, the methodologies that exist to lead the projects of tomorrow and the approach that companies could follow to be able to promote and foster cultural change within of their organizations towards a less traditional and more innovative culture.

This book is divided into 5 chapters that describe in detail first of all general concepts of digital transformation, the importance of strategy and the benefits it brings in this new world where the company that does not transform and is more efficient does not survive . The second part describes more representative success stories of companies that are being digitally transformed and the direct benefits they have obtained by applying digital transformation strategies. In the third part, the value chain of the mining sector is described, stage by stage, giving general notions of the activities and processes that each stage comprises. In the fourth part of the book, it presents the general technologies and experiences of digitization in the mining industry and how applied technology has revolutionized and will revolutionize traditional mining, making it more competitive and efficient. Finally, in the fifth part, some general recommendations and conclusions are given.

Chapter 1: Industrial Digital Transformation in the XXI Century

Digital Transformation

Digital transformation is a concept, which implies the intervention of technologies to encourage an organization to develop in the sense of being more efficient and being more profitable. For this, to be a digital industry in any sector, planning, pilot tests, investment, implementation, a cultural change and high-level operational management are required. It is important to note that, to create this transformation in a traditional company, it will be complicated, but not impossible. The transformation will not be disruptive but incremental and will allow digital transformation in the medium term.

The most used technologies in digital transformation projects are IoT, Blockchain, Big Data, Cloud Computing, Artificial Intelligence and Machine Learning. In addition to including technologies in processes of change towards the digital transformation of industries, it must also be included and it is fundamental to change the way of thinking of employees and senior managers. If the corporate culture does not support change, it will be difficult for a company to instill new business processes and achieve digital enlightenment, and even more so in traditional companies such as mining. Shifting to a digitally transformed business often means breaking internal paradigms and engaging with customers differently.

It is not impossible to imagine that, at any given time, companies vary in their digital maturity, depending on the sector they are in, compromising the quality of their leadership and business planning, and the agility with which new business processes, both both face-to-face and internal - can be developed and deployed in this digitization gap.

The pandemic effect

What generated COVID-19 in our society is a social and economic upheaval that has provided a means to assess the role of digital transformation in business in a more urgent way. Thus, the most digitally mature companies have weathered the lockdown and recession better than those that were at an earlier stage of their digital transformation. Additionally, the pandemic has given those left in a pre-pandemic digital age the impetus to launch aggressive strategies and adopt more digital technology.

Why is digital transformation essential for all companies?

In general, we know that competition in the market stimulates innovation; there will always be new ways to make processes more efficient, and competition can jump one-step further than our business. Thus, we must always be aware of what happens in technological advances and their application in our industry to prevent other actors from being ahead of us.

Digital transformation is essential and necessary for all types of companies regardless of the sector where they are or the size of the company. A nice benefit of digital transformation is monitoring everything in real time, seeing the real performance of your processes, and monitoring your customers. If there is no digital transformation strategy in your companies, we are left behind and we could be left behind with respect to other competitors.

-One of the most difficult industries to transform, but that most needs it, is the mining and manufacturing industry

Mining, as well as traditional manufacturing have frequently remained isolated, so digital transformation breaks the pre-established processes to implement best practice analysis together. Taking an approach to improve the efficiency of working capital, helps to improve the speed of the exploitation of mining companies, increasing the quality of what is produced, having fewer personnel in inhospitable places and by generating adequate training in new technologies, companies Mining companies, for example, can use digital transformation to work more securely, with lower costs and achieve better results.

Strategy for digital transformation

The digital transformation strategy is a detailed plan of how the business will face the challenges that appear due to the interaction of the physical, digital and human. To develop a roadmap or "road map" to digitally transform businesses in the short, medium or long term, we have to be guided by what a business wants as a result and not by technology itself.

-One of the reasons why a business cannot create its digital potential is due to the lack of a strategy from its senior managers.

Seven steps will be mentioned that will allow you to create a digital strategy.

1. **Understand the reason for digital transformation**

Digital transformation can mean many things to many people. First, many managers believe that technology use is the raison d'être of digital transformation and why transformation lags behind after implementation. The first thing we must have to start with is to identify the business needs and objectives and start creating a strategy from that understanding.

Thus, the business objectives of 5 to 10 years in the future will help to fit the digital transformation strategy. To continue, it is necessary to have well identified the business cases that generate value in the company.

It is important to note that digital transformation is not only an initiative or a project, it is a joint effort that will position the company for the future.

2. Prepare for culture change

It is clear that the support of senior managers and their enthusiasm to carry out the Digital transformation of the company is a key factor for the cultural change in the organization. Technology is not the key to digital transformation but people, and that is why you should focus on them.

Naturally, there may be many people who are resistant to change and skepticism that new things will be better. That is why we must prepare ourselves to feel such rejection. A team mentality must be developed with certain key employees of the company to serve and help take the first step to be able to execute the digital transformation strategy.

Having a key team that understands the vision of the organization regarding digital transformation will be of great value for the company to achieve its commitment and see the fruits in the near future.

Be clear that the transformation process will have challenges throughout the This is why transformation leaders have to strengthen the culture that allows all workers to learn from mistakes and create success from them.

When the digital transformation strategy is implemented, keep in mind that the daily activities of the workers will be affected that is why the identification of those processes where they can be improved in efficiency, effectiveness and productivity is the most important part of the digital transformation.

3. Start small, but with strategy

Remember that digital transformation is a path, not an event at a certain time, and identifying proofs of concept from the beginning and how to work on digital transformation projects is essential. This will result in the stages of future initiatives and help digital transformation leaders sell the strategy at all levels.

There are two characteristics to identify those projects of great impact and that will lead to the adoption of the fastest digital strategy: First that the projects are quantifiable and second that they quickly accelerate the value of the process.

Proving that the return on investment in a key project is a short time allows generating expectations in all workers and the culture takes hold faster. The best thing is to work on initiatives that add value in less than six months, which are traditionally called "quick wins". These projects will give measurable results in a short time.

The first Digital transformation project initiatives are key to providing tangible results and ensuring the long-term success of the implemented strategy.

It is important to take the time to identify those projects that allow you to take the leap of digital transformation.

4: Map the implementation of technologies

If companies don't change company processes and cultures, workers will only work with expensive technologies and the real value will not be seen.

Technologies are important and those that will fundamentally help on the path of digital transformation are: Mobile technologies; IoT; Digital Twin; Robotics; Cloud or cloud; Artificial Intelligence and "Machine Learning"; Augmented Reality and Add-on Manufacturing.

The aforementioned technologies will help achieve the expected results of digital transformation initiatives that deliver quick results. To implement the aforementioned technologies, it is important to select suppliers well, taking into account long-term relationships and strategies. Having experienced partners and the right product will help reduce the time our digital initiatives generate value.

Developing a roadmap with technologies with great impact on company processes will provide the basis for short and long-term initiatives and the success of the company's digital transformation program will be essential.

5: Find experts and allies on your digital transformation path

When exploring opportunities with new vendors or vendors that you have already worked with and have new technologies, it is important to understand whether the technology offered and vendor support can be scalable. Second if the provider and the organization have the same vision on digital transformation. Third, if the supplier can support us in a long-term strategy. Fourth, understand how the provider will complement the technology that already exists in organizations. Fifth, if the supplier has the correct technology and experience for our specific cases in our industry. Sixth, the results obtained with similar initiatives and applications in similar companies.

We may have the most incredible technology but not the right provider to help us implement the initiatives. If the suppliers and partners do not have the necessary experience, we will have many problems in the implementation. Many companies already come with a lot of experience in these innovative projects because they have made mistakes in other companies and their knowledge of them will help in the digital transformation strategy.

We must look at the strengths we have in our company, and look for those suppliers and colleagues who reinforce these strengths and who understand our business. Having these suppliers that allow us to accelerate results and lead initiatives is critical in our results to digitally transform our industry.

6. Collect feedback and redefine strategy as required

By having a vision of the digital transformation strategy and having a team of transformation leaders, the strategy uses the digital initiatives, the roadmaps and the allies and suppliers for the existence of the digital transformation. Before proceeding with the implementation of real cases, it is important to establish performance indicators or KPIs for each project.

We have to make sure that everyone is aware of the responsibility that is required to carry out each project and have results as expected.

At the same time we have ensure we have a good feedback loop and lessons learned with key stakeholders to ensure that everyone is learning from the experience of deploying the digital transformation strategy.

New objectives sThey will see coming in each project, but it is normal because the technology that is implemented will be flexible and agile as appropriate. With the digital transformation we are not going to box ourselves in a single space.

Understand that agility will be one of the characteristics of companies that get involved in digital transformation projects. That is why the roadmap is of vital importance to verify and correct the expected objectives according to the results that we are obtaining.

7. Scale and transform

If you are already having results from the digital initiatives that you are implementing, communicate the results and the success that you are achieving will help to generate more collaboration between all the workers to continue taking forward the projects as identified in the roadmap.

While the digital transformation strategy is implemented, new connections will emerge between people, processes and products. In this way, other areas and sectors of the company will be convinced to use new and digital technologies since the experience reflected will help to motivate them. The scaling will be horizontal and vertical

Digital transformation looks different in each company, so each organization has to diagnose and see what the correct strategy will be to lead the digital transformation.

As a summary of this part of the book, it is that while the results are seen in the short term, it will help to accelerate and continue implementing our digital transformation strategy. As long as companies and industries continue to evolve, the digital transformation will make people and processes more efficient and productive, generating better competitiveness, differentiated products and customer satisfaction.

THE DIGITAL TRANSFORMATION STRATEGY

THE 7 STEPS TO CREATING THE PLAN

1.) UNDERSTAND WHY DIGITAL TRANSFORMATION

Digital transformation can mean many things to many people.

2.) PREPARE FOR CULTURE CHANGE

Technology is not the key to digital transformation but people

3.) START SMALL BUT WITH STRATEGY

Remember that digital transformation is a path, not an event in a certain time

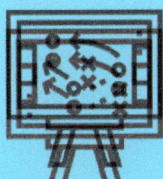

4.) MAP THE IMPLEMENTATION OF TECHNOLOGIES

If companies don't change company processes and cultures, workers will only work with expensive technologies and the real value will not be seen.

5.) FIND EXPERTS AND ALLIES ON YOUR DIGITAL TRANSFORMATION PATH

We must look at the strengths we have in our company, and look for those suppliers and colleagues who reinforce these strengths and who understand our business.

6.) COLLECT FEEDBACK AND REDEFINE STRATEGY AS REQUIRED

Before proceeding with the implementation of real cases, it is important to establish performance indicators or KPIs for each project.

7.) SCALE AND TRANSFORM

If you are already having results from the digital initiatives that you are implementing, communicate the results and the success that you are achieving will help to generate more collaboration

WWW.INNOVAYAP.COM

What are the most important technologies for a digital transformation?

Based on the "World Economic Forum", it has identified 8 basic technologies that the largest number of companies are thinking of implementing to achieve digital transformation. Each of the technologies mentioned plays a role in the strategy.

1. Mobile Use

For more than two decades, mobile phones and even more so "Smartphones" have irreversibly changed our world. By increasing the power of connectivity, the powers in our hands have permanently created lives influenced by technologies.

In all existing companies there are many opportunities present with mobile technologies, especially those that are already being implemented with 5G connectivity capabilities. Not surprisingly, 81% of CEOs of manufacturing companies see mobile as a strategic component of their vision for their companies.

Mobile phones are fundamental tools for changing the technological game. Thus, the more powerful mobile phones become and evolve and interconnected with 5G technologies, they will see incredible effects in all industries. Other technologies such as augmented reality, real-time analysis, among other functions of great value for any industry can be integrated into mobiles.

2. The cloud

In discrete, process-oriented industries, they have been hesitant to adopt cloud technologies. Basically the reasons found are due to a security issue, integration challenges and others that allow business continuity.

The concerns that have been mentioned have been disappearing with the new technologies that have appeared. That is why the cloud is increasingly a key component for digital transformation initiatives in companies due to its great flexibility, omnipresence, and agility to work transversally in any company.

3. IoT (Internet of things)

The internet of things is bringing impressive visibility and interest in both finished products, services or processes. Many companies are updating all their systems with IoT for data analysis and productivity of situations in general.

Many companies are achieving their goals of digital transformation using this technology. This is improving their efficiencies, flexibility and quicker responses to the market and customer needs. Achieving important innovations both in its processes, its products and its services.

4. Digital Twin

Digital twins, or Digital Twin, are helping also to achieve the objectives of companies in digital transformation. These are virtual models that represent real systems, but in virtual environments. They can be products, processes and tasks and are used to understand and predict behaviors of physical systems that can be simple or complex.

With the Digital twins companies have a clear display of their products and operations. The use of these systems becomes more relevant with the integration of augmented reality, IoT, and design software for products or CAD engineering.

5. Robotics

Many companies are implementing smart robots and their use and implementation are expanding more and more every day. The use of robots goes hand in hand with repetitive tasks and complemented with technologies such as IoT, artificial intelligence and sensors in general.

Among this area, it is important to mention unmanned vehicles or drones, which are used for different tasks, such as reconnaissance, surveillance, delivery of "delivery" and others.

Robots are a key component that help operations increase efficiency and have great responsibilities for improving worker safety and integrity.

6. Artificial intelligence and machine learning

The area of artificial intelligence is an area that plans to grow exponentially every year and become worth more than $ 17 billion in the next 5 years, according to "Research and Markets."

Due to the increase in the amount of information that is generated day by day in our lives and that with the different devices such as sensors or other tools that help us capture, artificial intelligence and "Machine Learning" are making data increasingly clear important useful that we could not see before. Thus as a result a new way of solving problems is emerging.

7. Augmented Reality (AR)

Augmented reality is the technology that allows connections between the digital world and the real world, with a single control.

Very well said what they describe to the RA as IoT for human beings. A clear example to better understand is how Microsoft workers use "HoloLens" to connect their world that surrounds them with the data that is generated and stored in the cloud, which performs a real-time analysis of the information that enter the Holo system. It allows the real-time visualization of the information.

The RA allows to improve the productivity of the workers, allows to have a better level of training for the workers and to improve the support services to the clients.

8. Additive Manufacturing

Commonly called 3D Printing, the process of building objects one layer at a time. According to McKinsey, the 3D printing industry will be worth $ 250 billion by 2025.

Industries seek to improve their efficiencies and using this technology generate great value throughout their production chain. In equipment maintenance, this technology makes it possible to manufacture components that could take time to reach the assembly site. This generates more efficiency, improves customer service, and reduces costs.

The eight technologies mentioned turn out to be the most common that can be seen in digital transformation strategies, but there are many more. To have success in the implementation of transformations, say it is not to implement a single technology, it is necessary to fully understand the course we are following and use the technologies as appropriate. A good understanding of how each of the technologies fits into our business is the key to success in our digital transformation program.

Benefits of implementing a digital transformation strategy

With a global perspective, understanding how companies do business, digital transformation must be approached with a strategy that combines expensive objectives where their competitiveness and strengths grow. The most representative benefits involved in digital transformation in companies are discussed below:

1. **Improve operational efficiency**

Technologies that involve automation and data analytics are improving operational efficiency by facilitating process change. Companies are now using connected technologies to gain visibility and control in their operations where timers and speculation existed only in the past. By applying the technologies correctly, inefficiencies can be identified and a company with a proactive strategy can develop agile processes that allow them to capitalize on innovation and the use of new technologies.

To close the idea, 39% of business and technology decision makers mention the cost reduction is due to a digital strategy.

2. **Better differentiation of products and services**

Differentiation is a critical factor for many industrial companies, especially manufacturers of products that often compete in crowded markets. Digital transformation can generate differentiation of product and service offerings by creating new opportunities and optimizing current delivery of products and services.

The important differentiation is shown in the time of the manufacture of the products, the commercialization, improvements of functionalities and improvement of the quality of the product. To achieve these goals, companies are leveraging technology to improve and better understand their processes and create improvements to differentiate themselves from the competition.

For example, real-time product simulations, changes and better collaboration can be done throughout the value chain from design to customer satisfaction tracking. Augmented reality is differentiating and enhancing value propositions and additive manufacturing is enabling new design possibilities, which means that technology is used to drive a state of constant change and adaptation.

Value propositions planned and carried out in technological environments allow companies to quickly recognize customer needs and go to market faster with solutions that had better meet customer requirements.

50% or more of B2B (Business to Buisness) companies will differentiate themselves by offering more personalized products and more connected shopping experiences.

3. Improved consumer experience

Organizations around the world are recognizing that focusing on the customer is critical to their digital transformation strategies and to effectively compete in the digital world.

Shoppers expect customizable products and experiences, so issues need to be addressed in real time or before challenges arise. The digitization of interfaces and communication with customers are becoming stronger and this has an impact on brand reputation.

Customer experience can now be measured in more meaningful and actionable ways than ever. At the core of this new normal customer success is the ability to create and create the most intimate relationships with customers through technological capabilities.

The construction of a culture focused on the customer where the cycles of interactions with customers can be improved with after-sales services where the performance of the implemented products is intelligently monitored, allowing a more proactive service and additional sales.

It can also take the form of new communication channels to achieve and accelerate customer time-to-market, such as guided maintenance and remote service enabled by monitoring of connected products.

71% of executives consider that they understand the impact that digital technology will have on customer behavior and preferences as their main challenge.

4: Create new business models

There is a massive shift in companies moving from independently selling products, to providing services beyond what they commonly manufacture or do. Business models change the traditional metrics of how revenue growth is identified by first evaluating recurring revenue and its evolution.

Traditionally, product monetization has ended at the point of sale; However, with the proliferation of digital capabilities, including cloud, mobile apps, and IoT, related services and revenues can help deliver that after-sales service. For industrial companies, this transformation of the business model creates an opportunity to offer new services and increase their profitability or not disappear from the competition.

30% of the industry's revenue will come from new business models by 2020.

5. Reduce business risk

Risk is more than ever a driver of change for companies in general. Between the shifting buyer-supplier dynamics we've discussed and the uncertainty of frequently changing global business and regulatory conditions, companies cannot afford to take high risks. In general, supply chains become more complex, and materials and goods move globally at a fast pace with prices varying at all times.

It is surprising to discover that many industrial companies still rely on heavy paper use to conduct business, where information can be difficult to interpret, maintain and disseminate, creating obstacles that increase risk.

Digital transformation allows for smarter decisions and lower risks, allowing time to take proactive action when there are big challenges.

Analytical capabilities and connected technologies are providing systems that enable real-time compliance reporting, such as plant risk assessments, forecasting systems, and quality inspection reviews on a global level.

Benefits of Implementing a Digital Transformation Strategy

Technologies involving automation and data analytics are improving operational efficiency by facilitating change in processes

Differentiation is a critical factor for many industrial companies, especially for manufacturers of products that often compete in competitive markets.

Organizations around the world are recognizing that focusing on the customer is critical to their digital transformation strategies

There is a massive shift in companies moving from independently selling products, to providing services beyond what they commonly manufacture or do.

Risk is more than ever an engine of change for businesses in general.

www.innovayap.com

Chapter 2: Success Stories of Digital Transformation Projects

Examples of digital transformation in the industry

The following case studies show how technologies are helping companies implement their transformation strategies. This information was provided by PTC, a company specialized in digital transformation worldwide.

Volvo case

Project Name: Reduce operating costs, improve quality, and improve worker productivity with information exchange.

The Volvo group turned to digital transformation as a differentiator for its trucks by maintaining standards of quality, flexibility and agility in its manufacturing processes.

To drive and improve your operational efficiency while improving product quality, a key differentiator was the implementation of a digital platform that encompasses information from systems in design, manufacturing, and quality control.

By synchronizing, Volvo was able to quickly adjust its production processes to keep up with changing customer requirements and customized to configured products. By ensuring that workforces in different roles continually access the latest information, they are driving transformation in the way they share and consume information throughout the value chain.

Lavihood case

Project name: Improve production performance and quality to effectively deal with the increasingly competitive food and beverage industry.

Lavifood is taking a fresh look at its production environments by integrating IIoT, industrial-grade connectivity, and real-time analytics at every level. Bringing plant-level intelligence to its production lines enabled this large agricultural company to provide more dynamic responses to potential problems, as well as reducing downtime and consequent impacts on products. This granular attention to detail is necessary for Lavifood to provide high-quality products that enhance the customer experience, as well as drive operational efficiency and product performance in its own plants.

Polaris case

Accelerate the introduction of new products and features to differentiate yourself from the competition

Polaris is using digital transformation as a means of differentiating its products and services from design to manufacturing to sales and marketing. The company has implemented additive manufacturing to improve the tool design process, creating more complex product parts faster and with less waste. Reducing design iterations and validation process time speeds time-to-market for new products and unlike the competition. By expanding out of the factory, Polaris is equipping its dealerships with live augmented reality experiences that illustrate potential product variations for customers.

Howden case

Reduce business risk and unplanned downtime

The global supplier of air and gas handling equipment is driving risk reduction and value for its customers through its digital transformation program called "Data-Driven Advantage."

By implementing various industry technologies to support its deployed products in customer operations, Howden is improving uptime and saving customers millions in unplanned downtime, substantially reducing business risk. Howdenis constantly improves the intelligence of its products and operations through digital twins. These twins further leverage artificial intelligence to predict failures and use the cloud as the supporting infrastructure.

Woodward case

Improve agility and intelligence to drive operational efficiency

Woodward, a leading provider of controllers and components to the industrial and aerospace markets, is executing a digital transformation strategy in its production environments to gain agility, save costs and improve decision making from previously isolated pockets of operational data in black boxes. The company is creating a "manufacturing information system" to synthesize this data; an accumulation of multiple pieces of existing digital technologies (ERP, CAD, PLM) and integrate IoT into your machines to get a live view of the operations of the entire plant. A key team for digital transformation, Woodward can now develop lenses basedin roles for different employees supported by the same plant digital twin. These lenses allow workers to respond to changes quickly and make more informed decisions.

Vodafone case

Creating an agile platform designed for better engagement

The main network operator Vodafone is expanding its value proposition to cities and municipalities through digital technologies. With IoT, the operator can differentiate their own customer-connected and correlated services to improve citizen engagement, cost management, sustainability, and economic prosperity. With its connectivity and the IoT platform backbone, Vodafone can begin to implement next-generation smart city applications including smart parking and air quality monitoring, further improving operational efficiency at the city-wide scale.

In any of the cases in which companies are starting or building new digital transformation initiatives, it is important to align business needs where an axis that intersects are the desired results using the digital transformation technology analyzed. It often takes a combination of multiple technologies, as well as people-driven process transformation, to achieve true differentiation. Digital transformation initiatives do not often fail when companies take a holistic approach and are prepared with a strategic plan where all employees can support and drive it.

Global digital transformation statistics

Digital transformation is not a requirement within 5 years, it is a race against time that needs to be addressed by the managers of all companies. The most outstanding statistics are presented on why digital transformation is important in industries.

- **Investment increases to be relevant**

The investment of technologies worldwide represented more than 2 trillion dollars in 2019. More than 40% of technologies developed aim to generate digital transformation in companies.

- **Company managers are leading the change**

Digital transformation is happening top-down, led by senior executives. In general, the digital transformation strategy is led by 28% CIOs (Chief Innovation Officers) and 23% by CEOs (Cheif Executive Officers). Many companies require change at all levels to embrace digital transformation and top executives are buying and leading it.

- **It is a necessary competence**

More than 70% of companies already have a digital transformation strategy or are working on one. For large companies the mentioned number would be validated.

- **Optimism in the role of IoT and Artificial Intelligence**

Six out of 10 company executives believe that these two technologies will play a very important role in their organizations and in the digital strategy of the business. 68% of company executives believe that the futures of business will involve AI working collaboratively with people. This information management through AI will allow leaders to make better decisions, increase operational effectiveness and innovate faster.

- **People are key to digital transformation**

71% of the population believes that people are extremely important to deploy a digital transformation strategy. Technology is nothing without people and that is why senior executives must think about how it will affect workers at all levels, the digital transformation, in their daily tasks and the technologies they will use.

- **Tangible business benefits**

40% of executives believe that digital transformation will improve operational efficiency, 36% improve its impact on the market and 35% improve customer expectations. In addition, 60% believe that with the help of digital transformation they have created new business models.

Chapter 3: General concepts of Mining

By reading this chapter, we want any reader to be able to better understand what the standard work of a mine is like, the processes that are established throughout the value chain of mining activities.

The mining activity is a legacy of our evolution as human beings and that shows our innovation of thousands of years ago to achieve better living standards. The discovery of fire begins a substantial improvement in life and that led to being able to melt the minerals and allocate them to obtain our first weapons for our safety and protection.

The decline in value of the mining sector

By identifying the mining value chain, we can identify 7 stages, which are: exploration, mine construction, exploitation, concentration in concentrator plants, smelting, refining and commercialization.

Exploration

Exploration is the stage in which the existence of a mining deposit is determined and the reserve (quantity of mineral) and quality (mineral grade) are determined. In the exploration we have several sub-stages that include the study of the area, the geoscientific interpretation (geophysics and geology) and the exploratory drilling.

In the study of the area, different activities such as surface geology are carried out in order to choose the best areas where the probability of finding mineral is high. For this, tours are made, physical samples are collected, aerial photos are taken, maps are reviewed, satellite images are reviewed and other evidence is reviewed that can guarantee that an area has high potential. From this stage it is already seen that the techniques and tools that can be used lend themselves to having innovative components and advanced technology applied. Some technologies that are currently used and that are revolutionizing exploration methods worldwide will be detailed later.

The second activity carried out in the study of the area is geophysical prospecting. The Ministry of Development of the Spanish government indicates that geophysics is the science that scientifically studies the earth and its application to the search for natural resources and tends to leave a lower environmental impact in the search for resources. In geophysical prospecting, different methods are used supported by tools that have been developed thanks to innovation.

Among them the seismic method, gravimetry, magnetometry, magneto telluric and geochemistry are used. The seismic method measures the reflection and refraction of seismic waves, gravimetry the variation of the components of gravity, magnetometry the variation of the earth's magnetic field, the telluric magneto measures the resistance of the ground from electric and magnetic fields of nature; and geochemistry measures the presence of organic or inorganic elements that can give evidence of the existence of some natural resource.

In mining, it is important to highlight that Aero magnetometry and gravimetry allow determining the thickness of the sedimentary layer and that they are used in the first exploration phases and currently unmanned equipment such as drones are used to perform measurements in explorations. In the next chapter, we will look at current innovations at each stage of the mining value chain that make exploration easier.

In the geo-scientific exploration stage (geophysics and geology), after the collection of the initial data product of field geology, aerial photos, satellites, gravimetry, magnetometry, geochemistry and other studies, they are processed by specialists in geo sciences (geophysicists and geologists), who interpret the product of data processing and give recommendations on the location of areas for the drilling of exploratory wells. Today, data analysis with specialized software is changing the way information is processed, increasing speed and giving better analysis to make better decisions. There are more and more professionals specialized in "Data Analysis", "Big data", "Data Scientist", "Data Automation", others;

As the penultimate last stage we have exploratory drilling, it is where the existence of mineralization under the surface is verified. The holes can be made from different angles, where the material is broken thanks to the drill that advances against the rock and returns to the surface. The material that is drilled is collected on the surface and sent for laboratory analysis. To understand well the shape of the mineralization, 3 D images are created where a clear image of the miner is indicated in the drilled area. In this, a large amount of information is also collected, it has to be stored, processed and issued reports so that the top management of the company can make decisions. The data analysis in depth greatly to help simplify the work of geologists.

The last stage is economic viability where uncertainties and probabilities of reaching viable mineral deposits are calculated. The market supply and demand of the mineral found is evaluated, creating reports of the current demand for the resource and the projected future demand, in addition to sovereign risks, such as government stability, and general details of the location of the deposit, labor, accessibility of roads, energy, telecommunications and other factors, which allow evaluating whether the project is viable or not.

Developing a new mine is very risky and requires a lot of planning to reduce uncertainty and unexpected events that may arise. Innovation precisely allows experimenting and trying new things that will reduce the level of uncertainty and risks in the early stages of any project. The next chapter presents these innovations that are making it possible to reduce risks in the early stages of the value chain of a mining activity.

Construction

Once the solid business case is in place, where the initial profile for the exploitation model, pre-feasibility studies and finally feasibility have been made; the construction stage occurs. It should be noted that the studies must have a detailed description of the tonnage and grades of the deposit, minimum cut-off grade, development plan, mining method, transportation of inputs, investments, royalties, insurance, taxes, general expenses, unit costs per ton and others. All this will serve to find a cash flow and finally be clear about the profitability of the project.

Specify that a final feasibility study is not the only thing that is required. It is clear that the mining company must obtain the socio-environmental permits from the State where it will carry out operations. For the construction and preparation stage of the mine, it is based on the estimates that have been made in the feasibility stage. The beneficiation plant has to be built and the mine prepared for when it is ready for the exploitation and extraction of the mineral, the preparation of the mine itself depends on whether the exploitation is open pit or underground.

In the construction of the plant, having already defined the appropriate method to obtain the concentration of the required mineral, a lot of thought must be given to the type of technology to be implemented. Nowadays, the need to have integrated control centers is mandatory, since the benefits it has will help to make the project more profitable in general. Although the investment may be a bit high at the beginning, throughout the operation, the investment pays off in a few years and helps increase the profits of the business.

What does it mean to have an automated plant? First, savings in labor, second, better efficiency in mineral recovery using artificial intelligence and "Machine Learning", third, better security and fourth savings in maintenance. All this is in accordance with the latest technological practices to be able to use predictive analytics, machine learning, augmented reality, virtual reality, robots for maintenance, quality supervision drones and thus other technologies that we will see in detail later.

In the preparation of the mine, based on the pre-established method of exploitation, the way must be prepared until reaching the mineralized zone. In the case of open pit exploitation, it is required to carry out drainage works and access work to the deposit. In the case of an underground mine, development work is carried out to the mineral through horizontal tunnels called galleries, vertical tunnels for ventilation called chimneys, vertical tunnels to be able to remove the mineral or personnel called shafts or ramps that are tunnels in the shape of spiral that allow transit and communicate the different levels that may exist in a mine.

To carry out these preparatory work, autonomous equipment and high-precision drone equipment can currently be used which can help improve the productivity and preparation of the mine before exploitation itself. We will talk about the equipment that has the prospect of changing how these types of work are done in the mining of the future.

It is important to note that, in order to carry out the construction and preparation of the mine, important details must have been given, such as where to cover the electricity supply for the operation of the mine. Thus, the traditional thing is to get supplies from a nearby electrical transmission line integrated into the national electrical grid. What is usually done is to install high-voltage towers and a substation to be able to provide enough power to the mining operation. The cost to be able to do all this infrastructure is high and that is why future mining projects should think about using renewable energies that allow them to complement the required energy or give them 100% autonomy. Battery technology is highly developed and can be used in these realities.

Exploitation

For the exploitation, depending on the type of work done both in open pit and underground, concatenated activities are carried out that are necessary to complete the exploitation cycle. In open pit the cycle is established by drilling, blasting, loading and transportation. It is used in large mining and involves high production.

With regard to underground exploitation, the cycle it comprises begins with drilling, blasting, hauling and transport outside the mine. Underground mines are very versatile and the way they carry out an exploitation cycle depends on the magnitude of the operation, the capital that the company has to be able to acquire technology and the infrastructure itself.

This is a stage where technology can help reduce the risk that workers have in this activity and also increase the productivity of the mines. I remember a saying of a close person, who has seen different sectors in their professional development, mentions that how it is possible to continue having difficult working conditions in mining jobs, having great technological developments that can achieve remote exploitation with minimal presence of personnel in the mines. It is where we currently aim to arrive in a mining of the future, where we can digitize and automate the entire mining cycle in both underground and open pit mines.

Likewise, there are developments to be able to carry out continuous, automated and digitized mining, and even better being battery-powered equipment. It is a vision not in the long term but in the short term and we are not surprised that in less than ten years we will have fully automated mines working both the plant and the mine itself.

Process Plant

After having gone through the exploitation stage, where the mineral is removed from the earth, it is necessary to give it a treatment to increase its purity and quality, since this extracted mineral is not marketable or has industrial utility. It is necessary to go through a metallurgical treatment called concentration and where the ore grade per ton is increased.

The methods to achieve this are diverse and depend on the type of mineral, its structure, other elements present, hardness and other characteristics. The most used methods are by flotation and leaching. In both methods, the size of the rock must be reduced by physical methods to release metallic particles and subsequently, the metallic concentration must be made by physical and chemical methods.

In flotation, the crushing and grinding process begins, called comminution, and then it is followed by a granulometric selection process with hydrocyclones, then the mineral is floated in flotation cells, what is not mineral goes to the tailings and what has mineral goes through a process of solid and liquid separation through thickeners and filters. Finally, there is a pulp of concentrates that could already be commercialized or give it an additional added value in the refining and smelting process.

In leaching, the mineral that is on the surface is regularly concentrated in the form of oxides, where it goes through a crushing process, from there the mineral is stored in "Pads", which are like artificial hills of different levels and finally through a dripping process that is placed on top of these hills, the liquid absorbs the mineral and runs downwards and is concentrated in ponds where they will later be pumped to concentration plants where they will finally have an adequate level of concentration prior to the smelting and refining stage .

We must think that we need to be more efficient and better recover minerals with innovative technologies that are in our hands. We must get rid of the thought that current technologies are expensive and that there is no money to invest. It has been shown that, although the implementation and automation in plants, the return on investment in the short term is a reality. Implementing predictive systems, data cloud, analytics and artificial intelligence is a totally acceptable technology in the task of making plants more autonomous and we must implement them now. Not only if the plant is already operating, but also taking it into account in the feasibility study prior to starting a new mine.

Smelting

It is the process that leads to the separation of the impurities of the metals contained in the concentrates through a pyro-metallurgical process, where high temperatures transform the solids into liquids. The purity of the liquid concentrates of these metals can reach over 90%.

Refinement

In the refineries, the purity of the mineral is further increased, which may already be suitable for industrial use and transformation. Processes such as fire refining or electrowinning can be used.

Commercialization

Finally, the cycle and the heat chain of mineral marketing ends with the marketing of both concentrates and refined mineral. The concentrate can be sold to national and international refineries and the refined mineral can be sold at an industrial level, to a manufacturer of metal parts such as the automotive sector, the machinery sector in general and also high-value minerals such as gold or silver that They have already been refined and can be marketed to international financial entities or also to national governments worldwide.

For the commercialization stage, it is necessary to use more innovative transports that allow the delivery of the agreement to be completed on time and with the required expectations. There are very sophisticated transportation technologies today such as mining pipelines, but they involve high capital and infrastructure costs. The traditional ones such as dump trucks, railways and boats are other methods that must be innovated and updated. We will talk a bit about the need to be able to use more railroads powered by renewable energy as in Germany, or autonomous dump trucks that could help improve the transport of minerals worldwide.

Chapter 4: Mining 4.0

In the study of the World Economic Forum and Accenture of 2017, it was estimated that the benefit for the mining industry based on a development and implementation of a digital transformation strategy would make a potential benefit of 190 billion dollars in the period 2016-2025. Thus, the mining industry has included among the group of sectors great potential to increase productivity from the digitization of its assets, the processes of the value chain, the relationship with customers and suppliers and improvements in its labor force.

In general, digital transformation is frequently mentioned as one of the main concerns among most large mining companies, and over the years it has generated great expectations regarding its benefits, despite all this the general level of digitization of the industry is still low.

There are many exceptions and several cases of mining operations where a high level of digitization and automation of their processes has been achieved. The LKAB, Kiruna and Malmberget iron ore mines, located in northern Sweden, are operated with a combination of remotely controlled and fully automated equipment for drilling, blasting and transportation processes.

Full automation and electrification are central elements in future plans for deeper levels, where KLAB Development has been working closely with high-tech companies such as ABB, Epiroc and the Volvo group.

Smart sensors and monitoring systems are also generating large amounts of data, but the plants and the mine are not fully digitized, so the full potential of data analytics is not being used. Thus, the goodness of advanced analytics to gradually support and automate operational decision-making processes is not being exploited and it is a huge gap to have better productivity and security that is being wasted.

Remember that, in the mining value chain, to propose a mature digitization strategy, we need technologies that allow us to capture information throughout the value chain, store it, process it, issue reports in real time and all centralized in a data center. integrated control.

MINING VALUE CHAIN 4.0

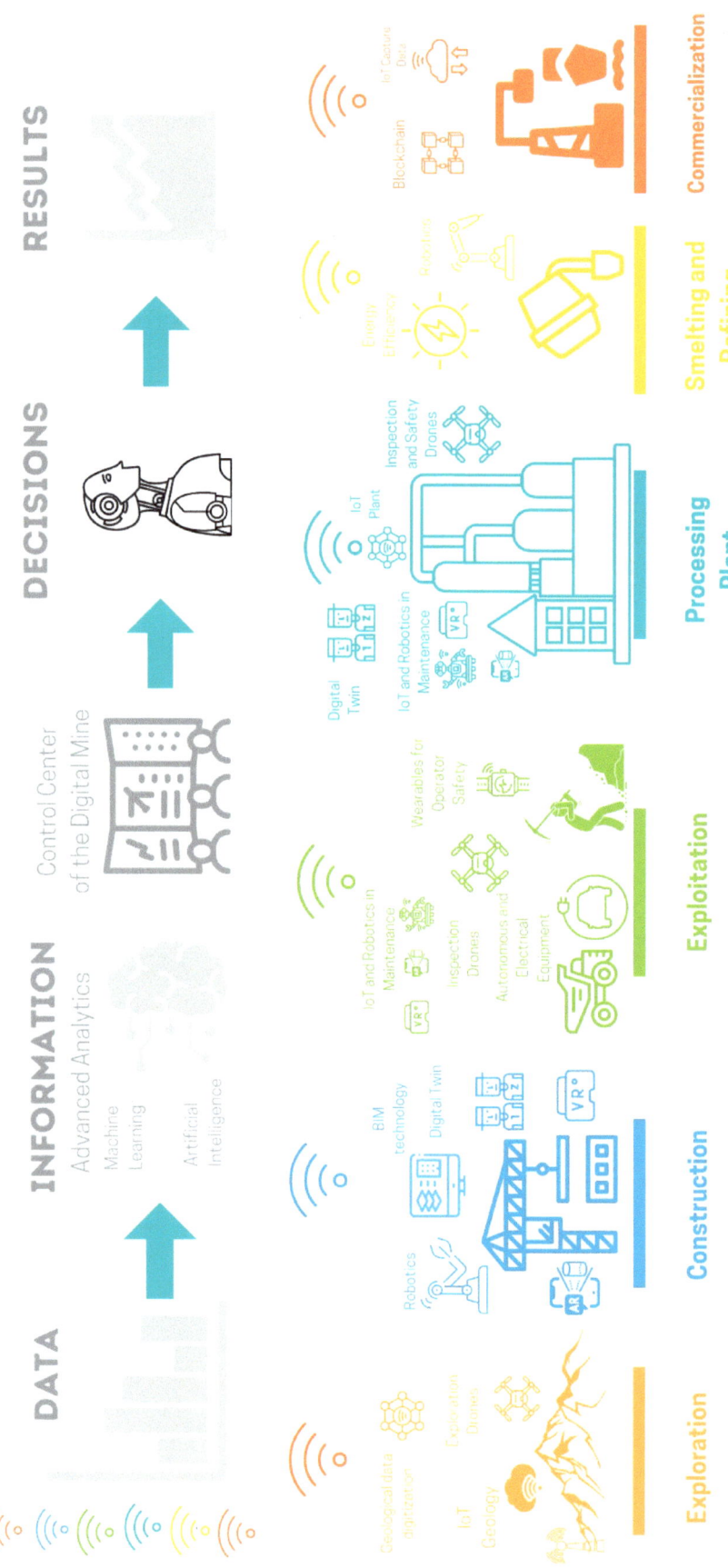

Mining 4.0: Exploration

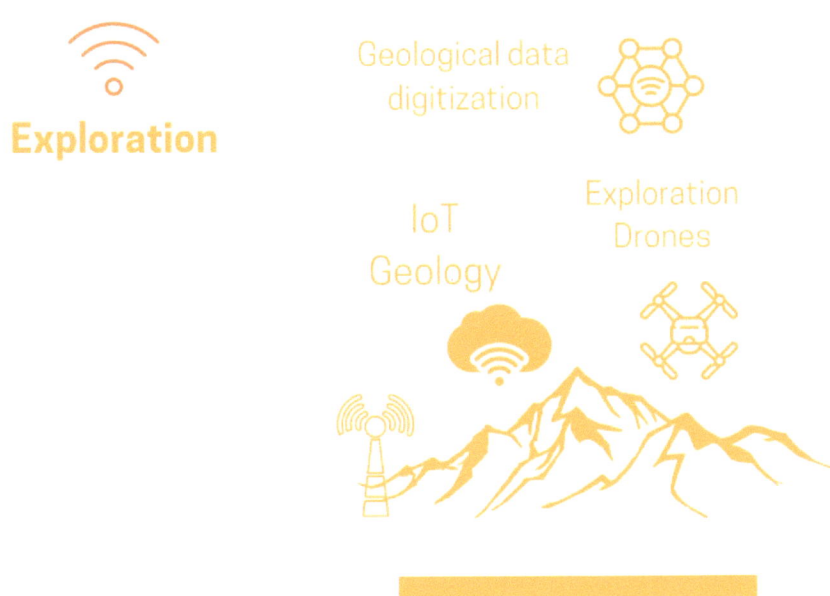

Digitization and processing of geological data with Artificial Intelligence

One of the most representative cases of digital transformation in the exploration area is what IBM and the giant Goldcorp are doing in alliance. The mining company gives the example of direction and strategy in exploration saying that they truly care about finding technological solutions that improve safety and the environment. And they look for strategic partners who also have the same vision.

According to the company, many gold deposits have already been found and mined, the easiest; on the other hand, discovering new mineral sources is a very time consuming, challenging and expensive process. More innovative ways to find inexpensive new gold mineralization are needed. In the mining industry, you must use your own assets more efficiently to be successful in your search

The company notes that IBM Watson and AI technology to extract more value from our data was the premise to take full advantage of Watson's machine learning and deep learning capabilities to predict gold mineralization, focusing on the Red Lake mines. Gold from Goldcorp.

The way to work with this technology is given in sequence; first, structured data is provided to Watson from various sources, such as 2D / 3D and drilling data. This information can be linked to unstructured data hidden on localized servers, speeding up geological analysis. Predictive models have been developed using machine learning to make predictions about gold mineralization and then they are interpreted by geologists.

Watson, the smart system, is not reinventing mining or replacing the institutional knowledge of the best geologists; what it is doing is the ability of the geologist. The IBM company says they have the technology and the people to drive innovation and Goldcorp has the knowledge and experience in geology and mining. Both companies are building an innovative product that will benefit and transform our industry.

Drones in exploration

Mining companies that use drones in their operations quickly realize the important added value that they provide at this stage of the mining value chain.

Surveillance, inspection, and surveying are just a few of the applications of drones in the field. For example, a drone can provide live visual support to a design or construction team, with outside teams viewing images. In general, the use of drones in mining is helping to improve the overall efficiency in search times for deposits, allowing more accurate and complete data that quickly details the site conditions.

Drone data can produce high-resolution ortho photos and DSM maps that would support exploration projects in areas where it is difficult to navigate on foot. Using drones costs only a fraction of the price of traditional manned aviation methods. And compared to land surveying team, it would take a team of surveyors weeks to collect the same amount of data that a drone can collect in a few hours.

IoT in geology

The Internet of Things is combining the physical world with the digital one, where everyday objects are installed with technology, such as sensors and Wi-Fi, to obtain a unique online identity and to be free to interact with the external environment.

IoT has great potential to revolutionize the area of geology by maintaining control over geological factors, so there are cases where the use of IoT allows to improve awareness of impacts, geological disturbances, in addition to helping geologists to improve their prospecting and identifying deposits in less time.

One of the important developments in intelligent systems is the portable spectrometer. This system allows different substances to be analyzed using different wavelengths. This technology allows to identify minerals in clays and silicates widely used as indicators of the presence of minerals such as gold, copper and uranium. This instrument allows doing this study in a short time and with much precision than just having the human eye.

Mining 4.0: Construction

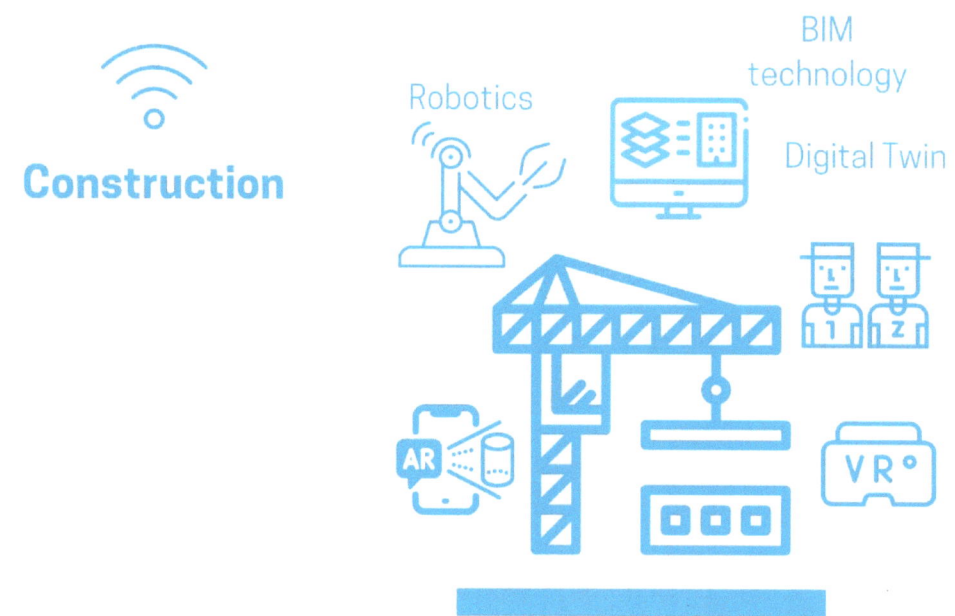

BIM technology

Innovative technologies are increasingly being used for the planning and construction of new or expansion projects of existing mining companies. One of them is the so-called "Building Information Modeling" or BIM, in Spanish information modeling in construction, defining it as a collaborative technology for the management of assets and construction projects in their different stages from design to construction.

This technology integrates people, processes and digital models for a management that unites engineering, construction and start-up operation. It is similar to a "Digital Twin" or digital twin where there are the essential building elements with 3D data of diagrams and elements, and also there are other levels of information such as time in 4D and costs in 5D.

In addition, the BIM system allows it to be integrated with other technologies such as Virtual Reality (VR) where engineers and workers can interact with the designed model and collaborate to improve each stage of the process from design to construction remotely.

The benefits that this technology is bringing is the reduction of risks of interference at the time of design, improving aspects such as maintainability and safety, optimizing work efficiency and efficiency to correct and improve the activities and work carried out.

Augmented Reality (AR) and Virtual Reality (VR) in construction

Augmented reality is fast becoming one of the must-have technologies for major industries. Thus, the mining industry is rapidly adopting it and it is allowing it to improve mine productivity at all levels, reducing maintenance costs, increasing the safety of mine workers with training prior to field activity, allowing training with sophisticated and modeled equipment, and execute construction projects with better efficiency since realities can be simulated and replicated and their progress evaluated.

AR technology provides a virtual overlay of information in a real world view. Construction workers equipped with AR headsets can view design schematics during on-site activities. The integration of portable devices with live video and sound transmission allows outside personnel to "see what the worker is seeing" - a technical specialist can provide live supervision and support to a team of workers on site.

Since mining is a very sensitive industry, the effectiveness of the processes to inspect the various problems that may arise and possible defects is critical in all processes. Remote assistance makes it possible for remote experts to inspect virtually all construction machinery and progress, as well as the surrounding environment. All this inspection is possible thanks to Augmented Reality, which integrated into the smart glasses through the zoom in / out function shows all the details, including dangerous areas, allowing to increase safety and the construction and commissioning of mines and plants profit.

Robotics in construction

Until recently, construction was one of the least digitized and automated industries in the world. Many projects could be completed more efficiently with the help of proper construction robotics, primarily because the related tasks are incredibly repetitive and stressful.

While manual labor is likely to always be an important component of modern construction, technology has been constantly improving since the first pulleys and power tools. Robots, drones, autonomous vehicles, 3D printing, and exoskeletons are starting to help get the job done.

Doxel Inc. has made a small tread robot that does exactly that: Scans and assesses the progress of a construction project across the site. The information you collect is used to detect potential errors and problems early.

Barcelona-based Scaled Robotics offers construction robotics that can be controlled remotely using mobile devices. The company's Husky unmanned ground vehicle can roam a construction site and capture critical information through multiple sensors. The data is transferred to the cloud, where it is used for building information modeling (BIM) of the project.

The global market for construction robotics also represents a great opportunity for developers and vendors. It could grow from $ 22.7 million in 2018 to $ 226 million in 2025, Tractica predicts. Research and Markets estimates that the market will grow to $ 126.4 million by 2025.

Digital twin in construction

A digital twin is a real-time digital representation of a physical object. Typically, digital data consists of sensors that continuously monitor changes in the environment and report up-to-date status in the form of measurements and images.

For construction, using digital twins means always having access to models as they were built and designed, which are constantly synchronized in real time. This allows companies to continuously monitor progress according to the schedule established in a BIM 4D model.

The difference between BIM technology is that this innovation digitizes engineering designs while digital twins or digital twins have the potential to really change workflows and have that reality at the right time to make the best decisions.

The concept of the digital twin, combined with portable and mobile devices on a construction site, can help to better represent the construction project at any time. It allows updated information to be fed back to the field to reduce the number of errors.

The applications of the digital twin in real time in construction sites, can give us the following benefits:

a. Automated progress monitoring

b. Comparison of As-Built versus As-Design models

c. Better resource planning and logistics

d. Better security monitoring

e. Better evaluation and quality control

f. Optimization of equipment use

g. Monitoring and tracking of workers

Mining 4.0: Exploitation

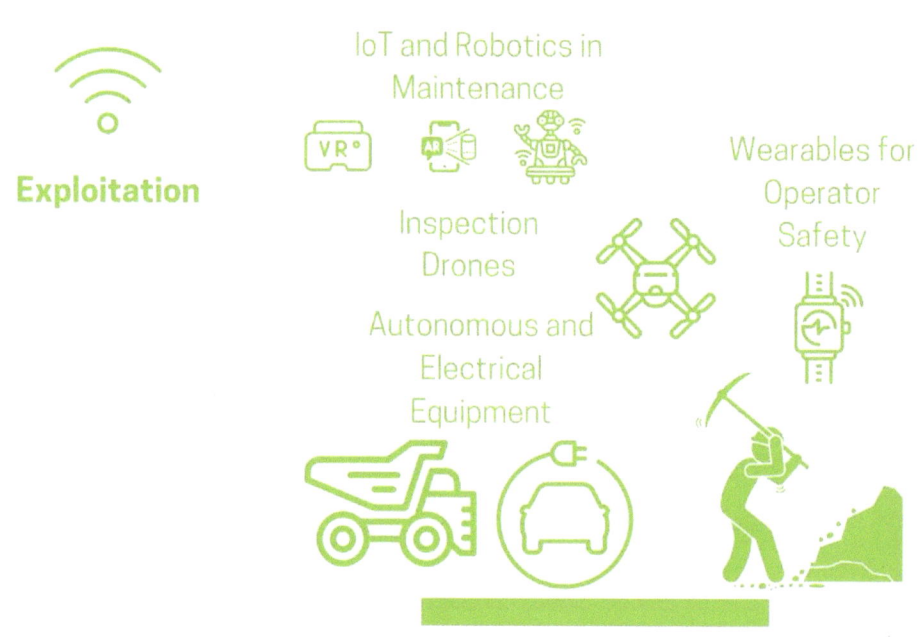

Autonomous Teams

We define autonomous teams as teams that do not require a person physically on the team or remotely to control its operation. Automated mining allows equipment to operate through a centrally controlled or scheduled routine, with few operation supervisors.

Autonomous teams must operate in a more predictable manner. With less direct operator control over trucks, drills, loaders, and other equipment, inherent operator-induced speed variations, material loading, ore travel and stripping, and downtime should be drastically reduced when using autonomous equipment.

Mining companies can experience cost reductions due to less wear and tear on equipment and fewer maintenance problems due to non-human intervention. Thus, the most important productivity indicators will show increases in the rate of availability and use of equipment. There will be greater productivity, since there will be no downtime during shift changes, as well as unplanned maintenance will be reduced because the planning and use of these equipment are scheduled in detail and there will be no mishaps.

ABB is already participating in the implementation of autonomous systems projects in mining, highlighting the levels of automation from those currently present in the industry, to those that are feasible in the future.

In order to define objectives in the transition to autonomous systems, it is important to establish a taxonomy so that automation providers and customers can define where they are and where they want to be in the short, medium and long term.

On the level 1, the systems provide operational assistance through decision support or remote assistance. Examples include software that helps locate underground mine vehicles.

Level 2 leans toward occasional autonomy in certain situations. Here, the automation system takes control in specific circumstances when and as requested by a human operator, for limited periods of time. People are still very involved, monitoring the state of operation and specifying targets for limited control situations.

At level 3, automated systems take over in certain situations. This can also be called "limited autonomy". A prerequisite is a complete and automated monitoring of the environment. An example would be autonomous drilling followed by autonomous loading of explosives in an underground mine.

At level 4, the system is in full control in certain situations and learns from your past actions, for example, so that it can better predict and solve problems on its own.

At the last level of this taxonomy is level 5, where a total autonomous operation occurs in all situations. No user interaction is required and humans can be completely absent. Today this is an aspiration, but, for example, an autonomous electric mining vehicle for full autonomous loading of ore would have significant safety and productivity advantages.

The use of autonomous equipment is allowing the reduction of events related to safety by eliminating dangers in the absence of operational personnel. Risks in the operation such as operator fatigue or decision errors when operating equipment could stop occurring.

In general, efficiencies in using autonomous equipment generate savings in working capital in existing mines and investment and operating capital in new mines that have a strategy of using autonomous equipment.

Electric vehicles in mining

The world is in the midst of a transition to a world powered by clean and green energy. The industry that will allow the change to clean energy will be the transportation industry, since it is a very important part of our life and without it we would not be able to develop as before.

Currently, the mining industry relies heavily on fossil fuels (especially diesel and electricity generated by coal) as a source of energy, especially for its daily operations. Therefore, energy costs are highly dependent on the price of the commodities of oil and coal. Thus, basically the gradual and rapid change in adoption of electric vehicles is accompanied by costs.

In the gradual but steady progression of the transition from fossil fuels to electric vehicles, we have also seen the large investment in R&D and incentives for the support structure, necessary to maintain a large population of electric vehicles. Fast chargers and the research that is being done is showing that there are possibilities to charge high-powered vehicles in just 3 minutes. This charge could be capable of producing enough energy to transport an electric vehicle more than 100 km.

The mining industry is looking for ways to save costs and reduce emissions, but the road to an all-electric mine is still a steep climb. As more mining trucks with an on-board electrical system enter the open pit mines, the electrification of haul roads is gaining ground. Thus, there are already mines where they have begun to make the leap to electrification using trolley lines to achieve it.

The advancement of the electric vehicle and plant industry is still in the early stages, with many of these vehicles still in the concept stages and, as with most emerging technologies, the only foreseeable way for a transition to occur. large-scale remains related to generating government incentives. It still takes a little more time, I would say 5 years, for the electric vehicle technology in mining to develop where there are more competitive prices.

ABB shows us in the following image, a scenario where all the mines should go and the electrification of trolley lines for transport trucks as an intermediate phase.

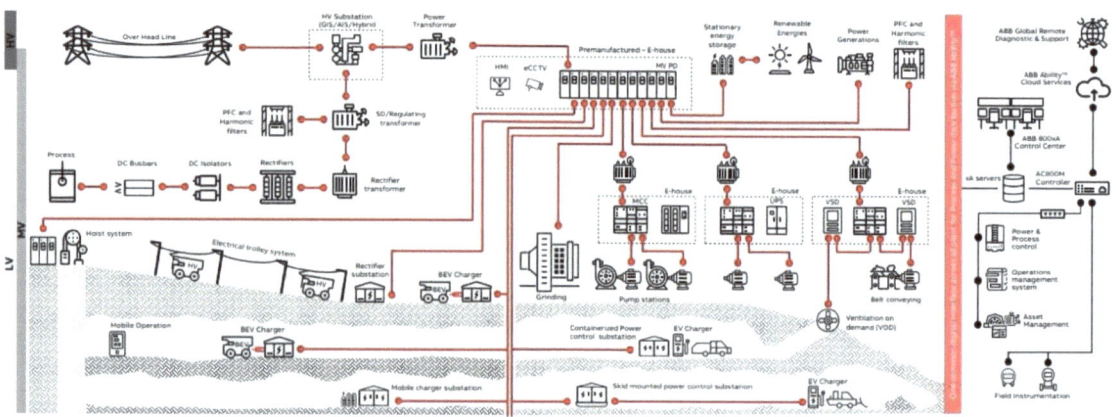

Source: A

Augmented Reality (AR) and Virtual (VR) in the Farm

Mining operations are usually quite difficult due to different environmental or climatic conditions. The help of augmented reality technology makes it possible for experts thousands of miles away to warn site operators about the potential dangers they may face. This real-time information improves collaboration and significantly increases response time allowing workers to be less exposed to various risks and more engaged with their tasks, increasing their productivity and work efficiency.

Assist by visual guidance using AR technology makes it possible to significantly increase drilling precision and efficiency, and which optimizes costs and improves efficiency at this stage of the exploitation process. By applying AR to aid drilling practices, it enables the transmission of necessary information by experts remotely.

Drones for Exploitation

The use of drones in exploitation is linked to production and security processes. For example, drones are currently being used in the inspection of tunnels in underground mines where due to the danger that could exist in certain spaces after the blasting, a drone can help to inspect and evaluate the conditions for the following works that can be done without endangering the lives of workers. Likewise, using drones in the exploitation allows us to ensure the quality of work in the mineral extraction processes and help planning engineers make the best decisions to continue producing mineral.

- **Stockpile management (Mineral stored in intermediate phases)**

Due to the irregular shape of stockpiles, it is difficult to estimate their volume with great precision using traditional methods such as GNSS surveying. These traditional methods also do not allow frequent surveys and impact on the safety of the surveyor.

Drones can be used to generate point clouds, digital surface models, digital terrain models, and a 3D reconstruction of a mining site. Since the point cloud contains several thousand data points, very precise volume calculations can now be easily performed.

Because the results of the drones and post-processing software are unbiased, another use you can make of the drone is to validate the amount of material moved by subcontractors.

- **Evaluation before and after drilling or blasting**

The use of drones in mining allows for affordable and cost-effective 3D reconstructions resulting in field surface models for the areas to be blown up or drilled.

These models help to accurately analyze the area to be drilled and calculate the volume to be extracted after blasting. This data allows you to better manage resources, such as the number of trucks required. A comparison with studies carried out before and after blasting will allow the volumes to be calculated with greater precision. In conclusion, the use of drones in exploitation improves planning for future blasting, optimizing the cost of explosives and time on site with drilling equipment.

Predictive maintenance in mobile operating equipment

With the era of Industry 4.0, it is no longer prudent, strategically or economically smart to wait until a critical mining asset has broken down and then repair equipment. Equipment breakdowns are often costly on many levels: downtime means productivity suffers, parts can be expensive, and then there are labor and energy costs.

Data analytics coupled with predictive maintenance can be a revolution and a boon for mining operations. Well, according to some experts, it would increase profits by increasing productivity approximately between 10% to 20%.

The evolution of maintenance has come a long way in the last decade, thanks to the availability of real-time data. Maintenance has 5 common approaches that can be applied to mine assets and that the help of digitization in these stages can go a long way in improving productivity and efficiency in maintenance. Steppes are reactive, preventive, condition-based, predictive, and prescriptive.

Moving from reactive to preventive maintenance will help improve asset reliability, but may not be effective as there will still be unplanned downtime and costly repairs that could have been avoided.

Condition-based monitoring, monitoring machines while they are still running, is the first step in adopting a forward-looking maintenance strategy. Data can be collected online through network connectivity to sensors or offline through operator rounds or other means, depending on the criticality of the machine.

Predictive maintenance further advances the condition-based approach through the use of model-based anomaly detection. It is based on the online collection of detection data and uses data analysis to predict the reliability of the machine.

The last level of maintenance, prescriptive maintenance, involves the integration of big data, analytics, machine learning, and artificial intelligence. Take predictive maintenance a step further by implementing an action to resolve an impending problem, rather than simply recommending an action.

The digital transformation strategy in maintenance will allow access to operational data in real time and reach the fifth level of maintenance, which is ideal for progressing in the strategic guidelines of any industry, being profitable and sustainable over time.

Wearable to improve safety and productivity

It is true that as the decrease in mineral reserves pushes the mines to deepen and to be in increasingly remote places, new technologies are being developed to reduce costs, risk and boost productivity. Thus, we can define Wearables as clothing, equipment and other electronic and intelligent accessories that workers use and that allow boosting productivity, but they are also fundamental to guarantee health, safety and improve the care of the environment in operations mining.

For a successful implementation, leaders must understand the business benefits, the impact on stakeholders, and the potential challenges that exist for implementation.

The different types that exist in the market and that generally include according to sectors, are the following:

Wearable environmental monitoring, where remote sensing technologies help frontline workers anticipate and companies respond to environmental risks.

Wearable focused on the health of workers, where biometric devices provide real-time alerts of physiological conditions of operators such as stress and also warn if they are at risk of injury or cause accidents.

Wearable training with adoption of new technology, where in a virtual way in real time it allows the training of new skills in workers.

Wearable of incidents in the mine, where monitoring and communication technologies are promoting the ability to alert and respond to incidents in real time.

Mining 4.0: Concentration and process plant

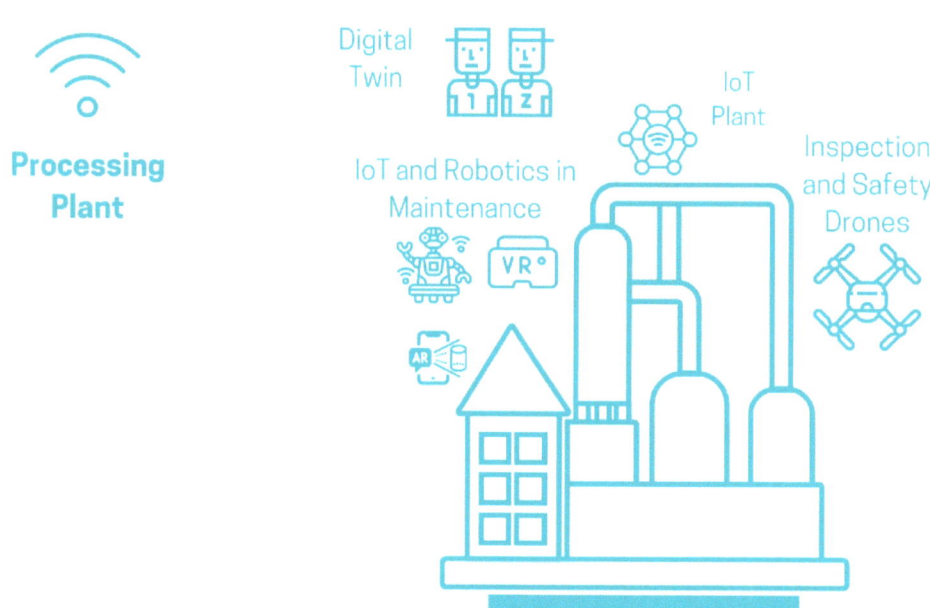

Digital twin at the mill

A Digital Twin for Mining is a virtual representation of the physical world, stored on a data platform in the cloud. Unlike older Programmable Logic Controllers (PLCs), Distributed Control Systems (DCS), and Mining Execution Systems (MES), Digital Twins use modern user interfaces (UI) and advanced visualizations to help the operator understand what what's happening in mine.

For example, the Digital Twin for Mining makes it possible to discover that a container in its crushing circuit is being overloaded and therefore the AI-predicted settings for the crusher width and feeder speed must be set. This will allow operators to make informed decisions with the ability to simulate and predict the downstream effects of those actions.

A Digital Twin for Mining integrates and extracts data from your existing dataset and industrial IoT sensors into a representative real-world structure within a platform.

How to design and implement a digital twin in mining?

The typical process for most mining operations when implementing a Digital Twin and Artificial Intelligence platform is:

1) Devise: Identify and quantify all the major opportunities within the mines and / or in your multiple mines. Define your preferred architectural strategy in advance

2) Define your strategy (broaden or deepen): understand if it is better to mention all your trades on the maturity curve at the same time, at a similar pace or if it is better to focus on part of your trades

3) Iterate - Deploy your solution on small incremental chucks to minimize risk, but also generate early profit. These benefits can support future projects.

4) Support: the general implementation process has to be supported with tools, a structure and systems of the project manager with specialized people to maximize success.

To reduce their costs structurally, mining companies must look to technologies that can change the mark in terms of reducing operating costs, optimizing processes and predictive maintenance. Thus, digital twins and artificial intelligence technologies have proven to be a success in the mining sector and are increasingly being adopted globally.

Predictive maintenance in plant equipment

Predictive Maintenance (PdM) allows you to predict equipment failures before they become a major failure and cause losses to mineral processing plants.

By gathering condition data from various assets through sensors located on valves, switches, motors, commutators, gearboxes, crushers, and conveyors, it enables all data to be collected, stored, and analyzed together. Thus, any specialized software can provide a warning of a potential failure (i.e., lubrication fails) with a proposed solution, all with enough time to address it before production is affected.

A critical analysis of each asset, equipment or component takes into account failure modes, available control system data, as well as information from pre-installed expert condition monitoring systems and data sheets.

There are four stages to implementing a comprehensive predictive maintenance strategy. During an on-site assessment, experts, along with site maintenance specialists, assess the precise maintenance needs. Then, a long-term solution is designed based on business goals and available technologies. In the next stage, the external technical team implements the online condition monitoring application. Finally, the same team with the help of plant personnel remotely measures and optimizes maintenance performance and presents the results in regular reports. Under a service contract, ABB ensures that the application is always up-to-date so that operations departments,

One of the big industries dedicated to evaluating predictive maintenance in mining plants is ABB. The ABB Ability ™ predictive maintenance service based on the ABB Ability ™ Asset Vista Condition Monitoring digital application for mining is offering plant operators in mines easy-to-use real-time reporting and reporting with a comprehensive overview of the status of their production assets.

Internet of Things (IoT) in the process plant

The Internet of Things (IoT) is the extension of Internet connectivity to physical devices and everyday objects. Integrated with electronics, Internet connectivity, and other forms of hardware, these smart devices allow themselves to communicate and interact with others over the Internet, and can be remotely monitored and controlled. In the mining industry, IoT is used as a means to achieve cost and productivity optimization, improving security measures.

In mineral processing plants, the greatest source of uncertainty is related to the properties of the mineral; plants must be able to react quickly to any change. The technology can track the movements and properties of materials from the mine to the processing plant. By using IoT in the entire process chain that happens in the plant, it will allow to track the movements of the mineral throughout the plant system, and within the most common processes in mining include crushing, grinding, flotation, thickening. The result is higher equipment utilization, higher production, and lower energy consumption.

The image below is a general diagram, proposed by Siemens, of the main processes that occur in a mine from extraction, processing and marketing. You can see the different sensors that can be found throughout this path including gas flow, level, weight, valve position, pressure, process protection, temperature and identification sensors. It gives us an important image of the number of elements that being intelligent can send data and control processes to maximize productivity.

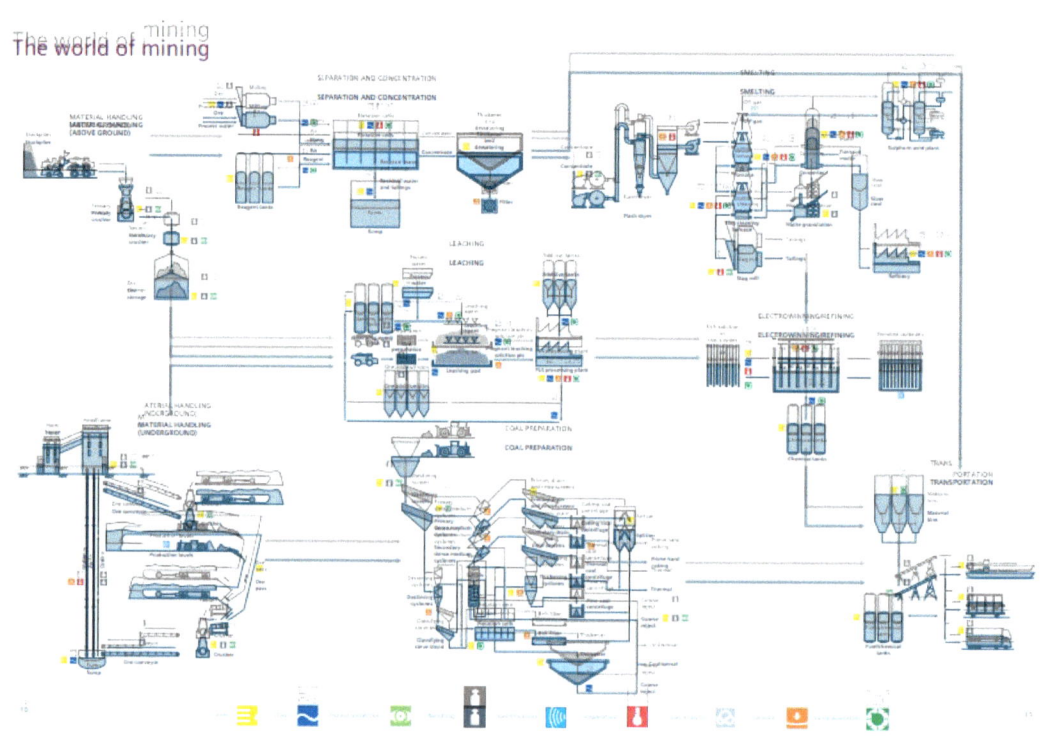

Source: Siemens

The main processes in a plant are described below and how the IoT helps to improve efficiency in each of these processes: Crushing, Grinding, Flotation and Thickening:

Integration of IoT in crushers will help save operating time and will be beneficial in detecting equipment failures remotely and helps to connect these devices directly with the help of plant managers in easy monitoring. With the help of IoT, the performance of crushers can be easily monitored by integrating remote monitoring systems with GPS (global positioning systems). This integration will ensure the real-time update of data, such as fuel efficiency malfunctions, assigning a series of activities in advance, and operating mobile crushers from remote locations.

A grinding circuit is a complex and multi-variable interaction system. Dynamically changing ore conditions and wear parameters pose particular problems for mills, which is why the IoT helps solve them. The data that can be measured from the grinding system allows a system to adjust certain parameters and optimize the process

The use of IoT in the float stage can lead to a significant improvement in the performance of this process. The data that is collected and through a central system will allow to stabilize the process and maximize the production of concentrate, while guaranteeing a minimum quality of the concentrated material.

Thickening is a process in which a suspension or a solid-liquid mixture separates into a thick suspension containing most of the solids and an overflow of essentially clear water. The driving force for separation is gravitational, where differences in phase densities drive the separation of solids and liquids. In mining applications, sedimentation thickening is applied to both product and tailings streams to recover water. This water is recycled in the process.

Having equipped with IoT in the thickening process allows to monitor the speed limitation variable and apply control strategies to maintain optimal conditions. Measurable parameters that can be used for thickener control include thickener feed rate, feed density, underflow density, overflow clarity, bed level, bed mass, tilt torque, rake height, solids settling rate, and underflow technology.

The industrial Internet of things is making it possible to cover all types of machinery with sensors connected to the Internet and this has generated that mining workers see the benefits of this. The most representative advantages of having IoT in a mining process plant are rescued below:

1. Time saving

When it comes to developing and maintaining equipment in a plant, the IoT helps dramatically. The data collected and a pertinent analysis save a lot of time in the maintenance and start-up of the equipment.

2. IoT improves security

There are still significant dangers in the mining industry, although technology has thus far managed to eliminate some of them. IoT can help prevent the collapse of unstable shafts, for example, because the sensors will collect data in real time and predict faulty equipment before problems occur.

3. Advances in mining automation

By having separate products - that is, vehicles and equipment that work together - more data is collected, which can improve accuracy rates. IoT also enables mining companies to find the latest innovations, technology, and development trends to stay on top of the market. Combined, these create a seamless experience, making work easier overall.

4. Predictive maintenance

Having a fully integrated network, monitoring every aspect of an operation becomes much easier and only leads to increased productivity and security. This includes detecting wear on vital pieces of equipment, as well as projecting when repairs or maintenance are required.

5. Energy and cost benefits

Investing in IoT reduces energy expenditure and maintenance costs for mining companies. By having a transparent system, where all parts are monitored, it allows a much more efficient process in terms of energy and cost savings.

Mining 4.0: Smelting and Refining

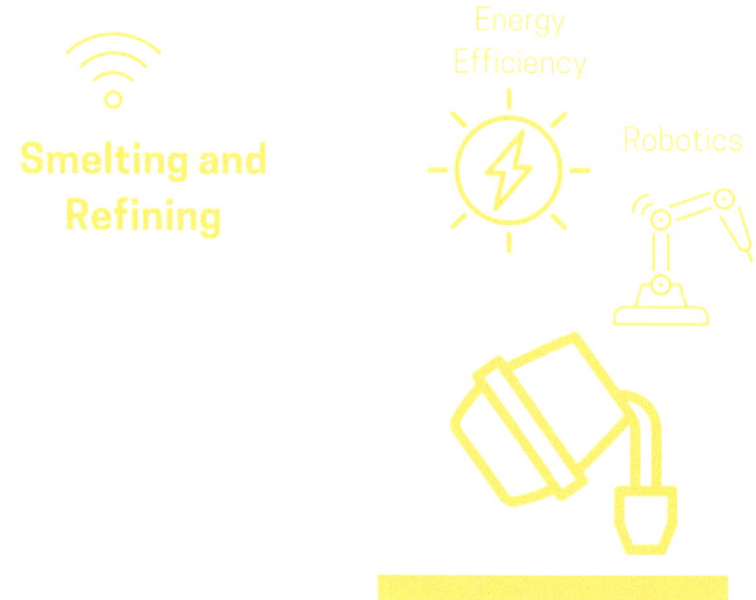

In metal smelting and refining, valuable components are separated from worthless material in a series of different physical and chemical reactions. The end product is a metal that contains controlled amounts of impurities. Primary smelting and refining produces metals directly from ore concentrates, while secondary smelting and refining produce metals from scrap and process waste.

Two metal recovery technologies are generally used to produce refined metals, these are pyrometallurgical and hydrometallurgical. Pyrometallurgical processes use heat to separate the desired metals from other materials. These processes use differences between oxidation potentials, melting points, vapor pressures, densities and / or miscibility of the mineral components when they melt.

Hydrometallurgical technologies differ from pyrometallurgical processes in that desired metals are separated from other materials using techniques that take advantage of differences in constituent solubilities and / or electrochemical properties while in aqueous solutions.

Using robots

Specialized Analysis Engineering, Inc. (SAE) is an engineering services company with expertise in industrial automation, robotics, process controls, systems integration, manufacturing, and wiring. The Logan, Utah-based firm has completed projects in a broad spectrum of industries, including arc furnace threading automation for phosphorous reduction furnaces and steel smelting furnaces.

Taking the aforementioned experience SAE has built a robot that is protected by a heat resistant armor manufactured by Roboworld, to handle high ambient temperatures and the extreme process environment that occur in the smelting and refining stage. SAE engineers worked with Roboworld to customize their robot armor to fit the inverted mounting position and required accessory equipment.

SAE also equipped the robot with custom-made, thermostatically controlled, electric resistance heaters on all joint motors and bearing housings to maintain operability during extremely cold winter months. They also installed auxiliary insulation in the robot's wiring harness under the protective suit. All conduit and wiring in electrical cabinets and operator panels in exposed locations were shielded and insulated to protect them from the harsh operating environment.

Robotics integrated into the company's foundry operations eliminates the possibility of human interaction with unsafe material handling, thus reducing the risk of injury to its operators. Robotic cranes help to stack refined metal packages and label them.

Better energy efficiency

The Swedish steel industry is facing a possible transition to drastically reduce its CO2 emissions through the direct use of hydrogen rather than continuing with coal-fired blast furnaces. The use of technological innovation to reduce the barriers to a transition and direct use of hydrogen in experimental work and to have the quantitative bases for its scaling up is being analyzed as a transition step.

Another success story is about the UltraFlex company. It is a leading provider of induction heating solutions with over 40 years of experience in the industry, it has manufactured a wide range of induction melting equipment so far, including melting furnaces, induction power supplies, induction coils, water cooling systems and more. Induction melting equipment has already been used successfully in 4 key induction melting applications: metal melting, precious metal smelting and refining, investment casting and smelting. This technology is allowing significant energy savings in this important process in the mining value chain and allowing comprehensive technological development with digital indicators that allow us to see productivity.

Mining 4.0: Comercialization

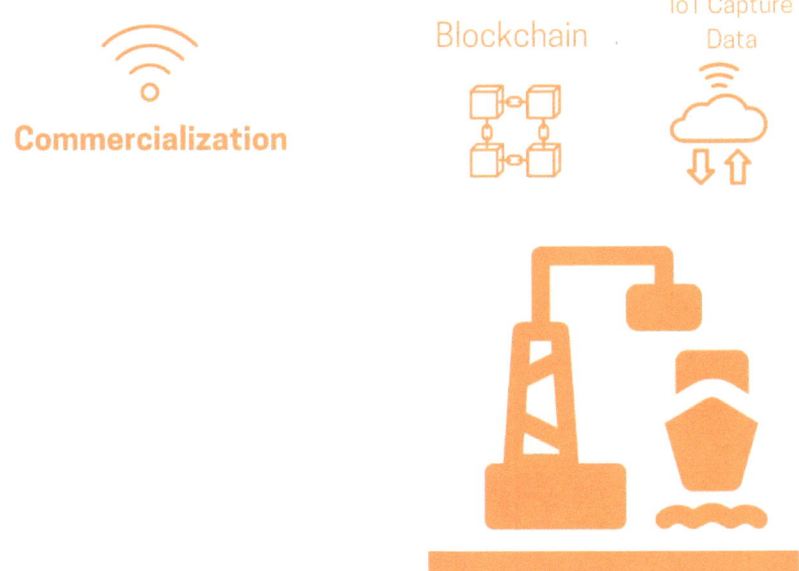

Blockchain

Blockchain is an innovative information registration system that avoids data manipulation. It is a kind of digital ledger of transactions to which all connected computer systems have access. It was originally created to ensure trust in financial transactions by creating unique digital elements that no one can change.

This new technology can be applied in other areas, such as mining, to track where materials come from, where they are produced, and whether or not they are sustainable. With the potential to improve supply chain transparency and traceability, the blockchain used in the mineral raw materials sector is therefore an ideal tool for building a responsible future, while reducing administrative costs.

There are three main aspects of mining in which blockchain technologies could provide added value and have a strong impact. First, blockchain can be used in mining site engineering, construction and handover, making transactions traceable during complex regulatory and standard management processes, and ensuring trust and labor compliance.

Second, blockchain is useful in the compliance and management of mining leases, which facilitates the workflow and visibility of documents, thus for example improving the traceability of the reserve estimate for stock market reports.

Finally, as mentioned above, the blockchain can be applied to the supply chain, in particular to track materials, from mineral blocks to concentrate to metal. This would improve transparency, clarifying the origin of the minerals and tracing all the steps to the end customer. In fact, many manufacturing companies are concerned about the origin of minerals and have decided not to use those that are extracted in conflict zones or sold by mining companies that do not comply with labor and environmental standards.

Therefore, although blockchain in mining can have several benefits, there are also limits to its use. Indeed, business processes can only be transformed if a number of conditions are met, such as the creation of a wide network of transactions and the willingness of the participants to provide accurate information on what and when and, in doing so, adhere to the basis of digital data, as the sole source of trust.

IoT in commercialization

An emblematic case that allows us to see how digital transformation is changing the way mining companies see commercialization in a more comprehensive way, is the iron mine located in Australia. This mine moved its production control, planning and short-term control to a new remote operations center.

The new operations center included the implementation and development of a supply chain visualization tool that looks at the iron ore supply chain as a whole and also displays important operational metrics in real time. The data is permanently displayed on large screens, with data coming from 16 disparate systems. This was the first time that the miner was able to see its entire supply chain in one place and that it has enabled decisions to be made for the entire business.

Digital Mine Control Center

With a centralized control room, you can monitor and control all equipment used in multiple mines, crushers, mills, foundries, mobile equipment, people, power plants, transportation information, and merchandising. It involves the recording and analysis of large amounts of processing variables, transmitted from locations miles away, enabling remote monitoring and advanced predictive and prescriptive analytics.

Control centers allow you to monitor, operate and control existing equipment and auxiliary facilities. Data, verbal information, radio signals, video images and satellite messages will be transmitted and collected centrally. Mine operators have a complete overview of mining sites at all times and can determine who else can receive the information. In addition to this, the control center will allow the information to be processed through data analytics, using tools such as Artificial Intelligence and Machine Learning to be able to generate relevant information for decision-making.

Digital Transformation in the Support Area

Digital transformation also occurs in mining support areas, so there are many technologies that can be used for example IoT in environmental areas to be able to ensure that environmental commitments are being met. Additionally, in the logistics area, blockchain can also be used to track purchases made by mining companies and there is no incorrect information from suppliers.

Human resources can use augmented and virtual reality technologies to train workers in specialized tasks such as operating high-tech equipment in operation. The Information Technology area must handle cybersecurity technologies in order to protect the company's data. The areas of logistics, Legal, communication and social responsibility can use RPAs (Robotic Process Autmation / Robotic Process Automation) to be able to help manage repetitive and tedious tasks in the mining company. Finally, the area of energy, innovation and maintenance will continue to use IoT systems to capture data from the processes that have relevance and allow them to transmit them to a centralized database at the corporate level that allows a quick and efficient use of information.

Do not forget that renewable energies are a topic of technological interest that is going to grow worldwide and the areas of energy, environment and innovation must keep them in mind for projects that reduce the carbon footprint of companies.

The Total Digital Mine

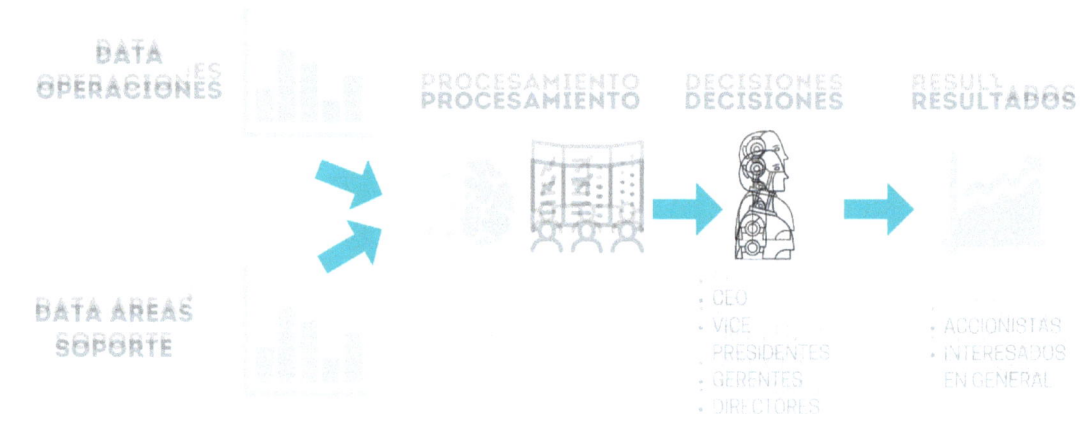

The digital mine will be a reality when the data of operations and the data of the support areas are integrated, they converge in a data processing system and finally the processed information can be displayed so that at different levels decisions can be made and the results generated that are aligned according to the strategic line of objectives of the mining company.

Chapter 5:
Conclusions and Recommendations

Conclusions and recommendations for a Mining 4.0 and a successful Digital Transformation strategy

Digital transformation plays an important role in the mining industry as the means to have a technified mining and the tool to improve the efficiency of its processes, reduce costs, and provide solutions to the growing social and environmental concerns among communities and local authorities. .

Technology has been crucial to allow the exploitation of new deposits that are becoming more complex every day to exploit, either due to lower ore grades than usual, more extreme climatic conditions, deeper deposits, harder rocks, hotter environments and more probability of accidents.

Mining companies that use digital transformation as a critical factor in improving labor productivity are more innovative. But there are still large gaps that indicate that there are low levels of intensity in R&D.

Although mining companies continue to receive support from manufacturers and suppliers with whom, working, and despite the fact that they offer them more innovative solutions every day, it is necessary to continue betting on new and collaborative alliances between mining companies, suppliers and research centers to continue developing technological innovations of great benefit to the mining industry.

There are several technology trends as major factors that will shape the mining of the present and the future. The digital transformation is the most relevant and allows in its implementation the incorporation of a set of tools called

4.0 technologies, to the mining business. Automation, robotics, Teleoperation, the internet of things, analytics, digital twin, BIM, drones and among others, have the potential to improve processes throughout the mining value chain.

Although digital transformation is the concern and path that many mining companies are following, the digitization of the industry in general remains low, indicating that most of the potential that digital transformation has has not been put to value.

The main challenges that companies must face to achieve a successful digital transformation is the strengthening of the innovation structure in the company and the commitment of senior leaders to prioritize a digital transformation strategy. In addition to this, the commitment and coordination of joint tasks between the different business units is required, implementing appropriate changes in the organizational structure and promoting a new culture of innovation and continuous improvement.

The great advances that would come in a 4.0 mining has to do with the electrification of the mine, circular economy, invisible mining and continuous mining. These concepts respond to the need to build a more sustainable and efficient industry, reducing the environmental footprint and improving the safety of mining operations.

Having a circular economy and using vehicles that do not run on fossil fuels is an "obligation" in a world that is moving away from these energy sources for cleaner ones. Every day more companies are evaluating the incorporation of electric fleets into their operations, since existing technologies can already offer economic alternatives, while R & D & I continues to advance in this matter.

Invisible mining using in situ leaching methods, have minimal impact on the surface and surroundings and generate virtually no waste. However, for a generalized application of this mining method, progress should be made in improving the permeability of the rock mass (for example, pre-conditioning techniques) and hydrogeological management, to ensure an optimal leaching process, in the first case, and minimize the risks associated with contamination of groundwater, in the second.

Finally, although the concept of continuous mining has been applied for many years in the coal mining industry, its application in other mining sectors has the potential to increase productivity, reduce costs and improve safety, along with the technological tools that brings the digital transformation regarding technologies such as automation, robotics and teleoperation both in plant and mine.

Bibliography

5 levels of automation for the autonomous mine of the future | ABB - Mining Automation and Integration | ABB (Mining solutions | ABB). (2020). ABB: https://new.abb.com/mining/mineoptimize/systems-solutions/mining-automation/5-levels-of-automation-for-the-autonomous-mine-of-the-future

A vision of all-electric mines is closer to reality than ever before - Mining electrification and infrastructure | ABB (Mining solutions | ABB). (2020). ABB: https://new.abb.com/mining/mineoptimize/systems-solutions/mining-electrification/a-vision-of-all-electric-mines-is-closer-to-reality-than-ever-before

ABB Ability Predictive Maintenance for mining - AssetVista - Digital mining services | ABB (Mining services | ABB). (2020). ABB: https://new.abb.com/mining/services/advanced-digital-services/predictive-maintenance-mining

Accenture. (2010). Using autonomous equipment to achieve high performance in the mining industry. Accenture Report. https://www.asirobots.com/wp-content/uploads/AccentureReport2010.pdf

Adams, T. (2020, March 15). Electric Vehicle Transition in the Mining Industry. GRT: https://globalroadtechnology.com/electric-vehicle-transition-in-the-mining-industry/

Adopting hydrogen direct reduction for the Swedish steel industry: A technological innovation system (TIS) study. (2020, January 1). ScienceDirect. https://www.sciencedirect.com/science/article/pii/S0959652619330550

Advanced Process Control and Analytics in industrial automation - What is new | ABB. (2020). ABB: https://new.abb.com/control-systems/features/advanced-process-control-and-analytics-in-industrial-automation

Advanced process control in mining | ABB - Digital applications for mining | ABB (MineOptimize). (2020). ABB: https://new.abb.com/mining/mineoptimize/digital-applications/advanced-process-control

All electric mine and trolley lines for haulage trucks | ABB Mining Tech Talks - Webinars for mining industry | ABB Mining Tech Talks. (2020). ABB: https://new.abb.com/mining/webinars/all-electric-mine-and-trolley-lines-for-haulage-trucks

Beginners guide to thickeners. (2020). Metso Outotec. https://www.mogroup.com/insights/blog/mining-and-metals/beginners-guide-to-thickeners/

Blockchain technologies and the mining industry: a shared future. (2020, December 17). European Economic and Social Committee.https://www.eesc.europa.eu/en/news-media/news/blockchain-technologies-and-mining-industry-shared-future

Deloitte. (2020). Future of mining with wearable.https://www2.deloitte.com/content/dam/Deloitte/global/Documents/Energy-and-Resources/gx-eri-norcat-report.pdf

Digital Twin for Mining. (2020). Insight Australia.https://au.insight.com/en_AU/what-we-do/digital-innovation/solutions/digital-twin-mining.html

Digital twin of material handling chain - ABB Ability Stockyard Management System (Mining operations and production management | ABB). (2020). ABB.https://new.abb.com/mining/mineoptimize/digital-applications/operations/abb-ability-stockyard-management-system/digital-twin-of-material-handling-chain

Digitalization develops jobs and companies | ABB. (2020). ABB.https://new.abb.com/mining/reference-stories/underground-stories/digitalization-develops-jobs-and-companies

Digitalization of Short Interval Control (SIC) and Production Scheduling in mining | ABB - ABB Ability Operations Management System for mining (Mining operations and production management | ABB). (2020). ABB.https://new.abb.com/mining/mineoptimize/digital-applications/operations/integrated-mine-operations/digitalization-of-short-interval-control-(sic)-and-production-scheduling-in-mining

Fell, R. (2019, November 27). How AI technology promises to aid in finding gold. Let's put smart to work. IBM Field Notes: Stories by our clients.https://www.ibm.com/blogs/client-voices/ai-technology-promises-to-aid-in-finding-gold/

Financialnewsmedia.com.(2020, January 14). How Augmented Reality is Disrupting the Mining Industry in 2020. FinalcialNewsMedia.https://www.prnewswire.com/news-releases/how-augmented-reality-is-disrupting-the-mining-industry-in-2020-300986253.html

Hibaieva, A. (2018, April 10). How Digital Twin Technology Can Help the Construction Industry | Intellectsoft US. Intellectsoft Blog.https://www.intellectsoft.net/blog/advanced-imaging-algorithms-for-digital-twin-reconstruction/

Insight.(2017, May). IoT, Digital Twin and AI in Mining. Insight Report.https://au.insight.com/content/dam/insight-web/en_AU/pdfs/insight/Insight%20Report%20-%20IoT%20Digital%20Twins%20and%20AI%20in%20Mining.pdf

Integrated Remote Operations Center - Central Control Room Mining | ABB - Open-pit mining. (2020). ABB. https://new.abb.com/mining/open-pit-mining/central-control-room

Malecaj, L. (2020, December 18). Augmented Reality Revolutionizing the Mining Industry. VSight Remote. https://www.vsight.io/augmented-reality-revolutionizing-the-mining-industry/#:%7E:text=Optimization%20of%20the%20mining%20process,benefits%20that%20this%20technology%20offers.

Manager, IC,(2020). Smelting and Refining Operations. Iloencyclopedia. https://www.iloencyclopaedia.org/part-xiii-12343/metal-processing-and-metal-working-industry/itemlist/category/135-smelting-and-refining-operations

Matthews, K. (2019, October 18). Construction robotics is changing the industry in these 5 ways. The Robot Report. https://www.therobotreport.com/construction-robotics-changing-industry/

Merry, H. (2019, February 18). 5 benefits IoT is having on the mining industry - Internet of Things blog. Business Operations. https://www.ibm.com/blogs/internet-of-things/mining-industry-benefits/

Model Predictive Control - MPC technology from ABB - What is new. (2020). ABB. https://new.abb.com/control-systems/features/model-predictive-control-mpc

More Industrial Automation, Robots and Unmanned Vehicles Resources. (2020). Robotic Tomorrow. https://www.roboticstomorrow.com/content.php?post=15879

Perroud, D. (2021, January 29). Mining and ... Wingtra. https://wingtra.com/drone-mapping-applications/mining-and-aggregates/

Report L ..(2020, July 9). Autonomous Mining Equipment Global Market Report 2020-30: Covid 19 Growth and Change. GlobeNewswire News Room. https://www.globenewswire.com/news-release/2020/07/09/2060321/0/en/Autonomous-Mining-Equipment-Global-Market-Report-2020-30-Covid-19-Growth-and-Change.html

Salamanca, H. (2005, November 10). Robot system and method for scrap bundling in metal smelting and refining processes. Google Patent. https://patents.google.com/patent/US20070299556A1/en

Sánchez, F., & Hartlieb, P. (2020, July 23). Innovation in the Mining Industry: Technological Trends and a Case Study of the Challenges of Disruptive Innovation. Mining, Metallurgy & Exploration. https://link.springer.com/article/10.1007/s42461-020-00262-1?error=cookies_not_supported&code=a8367097-2fb7-4360-a55a-6b12f2230f01

Stutt, A. (2020, April 18). A guide to predictive maintenance for the smart mine. MINING.COM. https://www.mining.com/a-guide-to-predictive-maintenance-for-the-smart-mine/

The smartest place I know: "This will be the decade of the digital twin". (2020, June 17). The Possible. https://www.the-possible.com/digital-twin-in-construction-industry/

Vavra, C. (2019a, May 31). Control Engineering | Robots developed and designed for hazardous, extreme conditions. Control Engineering. https://www.controleng.com/articles/robots-developed-and-designed-for-hazardous-extreme-conditions/

Vavra, C. (2019b, August 27). Control Engineering | Industry 4.0, automation advance smelter technology. Control Engineering. https://www.controleng.com/articles/industry-4-0-automation-advance-smelter-technology/

White, N. (2020, November 30). 5 Digital Transformation Examples in the Industrial Enterprise. PTC. https://www.ptc.com/en/blogs/corporate/digital-transformation-examples-enterprise